The Ultimate DIY Solar HWS Enthusiast's Guide

By Jonathan Warnes

Dedication

Dedicated to my father Peter Warnes, engineer, with whom none of this could have been possible.

Table Of Content

Dedication 2

Preface 1

TANK 3

System Pressure and Pump Requirements 4

Choosing the Right Pump 5

Preventing Reverse Thermosiphoning with a One-Way Valve 6

Why is a One-Way Valve Needed? 6

How the One-Way Valve Fixes This Issue 6

My Experience with This Issue 6

Choosing the Right One-Way Valve 8

Installing the Circulating Pipe System 9

Important Safety Steps: 9

Connecting to the Tank: 9

Pro Tip: Use Plumbing Sealing Compound Instead of Tape 10

Plumbing the Circulating System 11

Connecting the Circulating Pump to the Bottom of the Tank 11

Connecting the Return Line to the Top of the Tank 11

Safety Considerations: The Pressure Relief Valve 12

Never Remove the Safety Pressure Valve! 12

Why the Valve Releases Water Normally 12

Alternative Tank Connection Options 13

If Your Tank Has Extra Ports 13

If You Can't Access the Bottom of the Tank 13

If You Can't Access the Top of the Tank 13

Choosing the Right Solar Collector ______ *14*
Finding the Right Size and Type ______ 14
Installation and Placement ______ 14
Sizing for Your Household ______ 15
Temperature Monitoring and Pump Control ______ *16*
How the Controller Works ______ 16
Setting the Differential Temperature ______ 17
Insulating Your Pipes for Maximum Efficiency ______ 17
Collector Temperature Tracking Example ______ *19*
Controlling the Pump ______ *20*
1. Simple Solar-Powered System (No Controller) ______ 20
2. Using a Pre-existing or Second-hand Controller ______ 21
Controller Unit – The Arduino Solution ______ *22*
Pump Speed Control and Noise Reduction ______ *24*
How to Implement Pump Speed Control: ______ 24
Optional Redundancy for Pump Failure: ______ 24
Pump Failure and Overheating: ______ 25
System Programming: ______ 25
Using DS18B20 Temperature Sensors with Arduino ______ *26*
RTC (Real-Time Clock) Removal: ______ 26
Setting Up the Arduino and LCD: ______ 26
Button Controls: ______ 27
Wiring and DS18B20 Setup: ______ 27
Wiring Up and Installing Temperature Sensors ______ *29*
Sensor Installation Methods: ______ 29
Useful Resources: ______ 30
Wiring and Configuration ______ *31*

LCD Wiring: ____ **31**

Dump Solenoid (Overheat Protection): ____ **31**

Code Snippet: ____ **32**

Safety Considerations: ____ **32**

Disclaimer: ____ ***48***

Important: ____ **48**

Preface

This guide is for homeowners with some technical, electrical, and plumbing skills who want to safely construct and use a solar hot water system in their own homes.

The system described here is based on the one I currently use, which has achieved a payback period of just one year. In contrast, most commercial solar hot water providers claim payback periods of 3 to 6 years, with one even suggesting that 10 years is reasonable.

This DIY system offers significant cost savings by utilizing an existing hot water tank. If your current tank is no longer serviceable, many perfectly good electric hot water systems can be obtained for little to no cost. Homeowners frequently discard them when upgrading to commercial solar systems, making them an excellent resource for this project.

Before installing this system, I contacted several local solar hot water providers for quotes and was disappointed by the cost and value they offered. One company quoted me $5,500—even after applying the local and federal green energy rebates! Another provider offered a system for $3,500, but it was a small unit that appeared to use lower-quality components.

At first, I was determined to find a system that took advantage of the rebates to reduce costs. However, I realized that even after applying the rebates, the payback period (the time it takes for energy savings to cover the cost of installation) was still five or more years.

This led me to explore a DIY approach. With some careful planning, research, and hands-on work, I was able to achieve a much better result—despite not having access to government rebates.

After using my system for about 15 months, it appears to have already paid for itself—essentially within one year! My total upfront cash cost was around $950, though I already had some parts on hand and used my existing hot water tank.

Before installing the solar hot water system, my 3 monthly water heating bills, including the service connection fee, were about $145. My most recent bill showed my hot water tariff cost at just

$25—during the worst quarter (winter). In spring and summer, my hot water costs are nearly zero!

TANK

I have found that using the existing hot water tank presents no major issues, except that if the tank is quite old, it may only last a few more years before developing leaks. If possible, replacing the inner cathodic protection element during the upgrade is a wise move, as this can help extend the tank's lifespan.

One major advantage of using the existing tank is that you can retain the built-in electric heating element as a booster for cold, cloudy, or rainy weather. I have my heating element set to a temperature of around 40°C. This means that if the solar heating does not fully warm the water during the day, the booster will activate during the off-peak night rate period—only if the water temperature drops below 40°C. It then heats the water up to approximately 50°C before shutting off.

Most hot water system thermostats are adjustable. Before making any changes, note the current setting—typically around 60–70°C. Once the solar hot water collector is installed, you can lower this setting to reduce electricity use and start saving on your water heating bills.

In my system, under normal weather conditions, the electric booster rarely activates because the water in the tank remains above 50°C, even after everyone in the household has taken their evening showers. During the day, the sun heats the water up to around 70°C.

Since solar hot water systems can generate extremely hot water under certain conditions, it's important to ensure safety at the tap. Many regions now require a tempering valve to be installed on any new system, though regulations vary by state and country. A tempering valve mixes cold water with the hot water from the tank, reducing the temperature before it reaches your taps. Be sure to check your local regulations for specific requirements. Keep in mind that water at just 50°C can cause serious burns, especially to infants.

System Pressure and Pump Requirements

In my installation, the hot water tank is ground-based, meaning the entire system—including the storage tank, circulating pipes, valves, collector, and connections—is all at mains pressure.

The pump's role is simply to circulate water, transferring heat from the solar collector to the tank. There is no "head" pressure in the system because, as the pump pushes water up, gravity simultaneously pulls water down from the collector, balancing the forces. This allows the system to operate efficiently with a small, low-power pump.

A 10 to 20-watt pump is sufficient. When I installed my system, I purchased a 15-watt pump for just $80, delivered to my door.

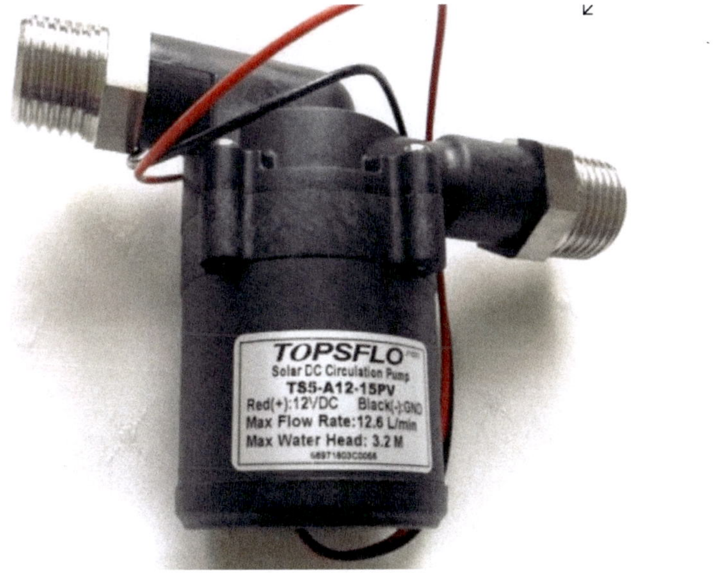

Choosing the Right Pump

- **Mains Pressure Rated** – The pump must be designed for mains pressure operation.
- **Heat-Resistant** – The pump must withstand exposure to high temperatures in case very hot water circulates back into the system.
- **Avoid Low-Quality Circulating Pumps** – Cheap, low-pressure pumps can fail under heat and pressure, potentially causing severe leaks and damage to your plumbing.

Preventing Reverse Thermosiphoning with a One-Way Valve

To ensure efficient operation, a one-way valve (non-return valve) should be installed just before or after the pump. This prevents a phenomenon known as reverse thermosiphoning, which can occur during cooler times of the day, particularly at night.

Why is a One-Way Valve Needed?

At night, the temperature at the top of the tank may remain around 60°C, while the bottom of the tank can drop to 20–30°C. Because warm water expands slightly, it becomes less dense and lighter, creating a natural tendency for it to rise into the solar collector. Simultaneously, the cooler, denser water at the bottom of the tank sinks back down into the return pipe.

This unintended circulation happens in the reverse direction of the pump's normal operation, effectively allowing heat to escape from the tank, cooling the water instead of keeping it warm.

How the One-Way Valve Fixes This Issue

By placing a one-way valve in the correct direction of flow—after the pump—you ensure that water can only circulate when the pump is running, preventing heat loss through reverse thermosiphoning.

My Experience with This Issue

During initial testing of my system without a one-way valve, I noticed that:

- Warm water would flow into the collector overnight, causing it to heat up.
- The control unit would detect the rise in temperature and activate the pump to return the warm water to the tank.
- This cycle would repeat every 30 minutes, leading to unnecessary pump operation and energy loss.

Once I installed a one-way valve after the pump, the issue was completely resolved, ensuring proper system efficiency.

Choosing the Right One-Way Valve

Finding a suitable one-way valve isn't always easy. I found mine on eBay for around $10. The ideal valve should be a simple loose spring and seat design, which naturally stays closed but opens under very little pressure.

Since the circulating pump is low-pressure, the valve must open easily. Most standard one-way valves sold at plumbing stores—such as those used to control cold water entering a hot water system—are not suitable for this application.

If you're unsure, visit your local plumbing shop and ask specifically for a low-resistance one-way valve suited for low-pressure circulation systems.

Installing the Circulating Pipe System

To integrate the circulation system into your hot water setup, you will need to turn off and isolate the system before making any modifications.

Important Safety Steps:

1. Turn off the mains power to the hot water system and lock it off to prevent accidental reconnection.
2. Shut off the incoming mains water pressure and ensure that no one can turn it back on while you're working.

Connecting to the Tank:

I found that the lowest fitting on my hot water tank could be removed, allowing me to install a T-adapter to provide an additional connection point for the circulation system.

Pro Tip: Use Plumbing Sealing Compound Instead of Tape

When making threaded connections, I highly recommend using a high-quality plumbing sealing compound instead of sealing tape. Compound is:

- **More reliable** and ensures a leak-free seal.
- **Less likely to clog** valves or system components if any residue enters the plumbing.

By following these steps, you can ensure a **reliable and efficient** solar hot water system without common installation issues.

Plumbing the Circulating System

Connecting the Circulating Pump to the Bottom of the Tank

To ensure the pump **draws cold water from the bottom of the storage tank** and sends it to the **solar collector on the roof**, you need to create a connection at the **lowest point of the tank**:

1. **Install a T-Piece** – I removed an existing fitting at the bottom of the tank and installed a **T-adapter**.
2. **Add an Isolation Valve** – Onto this T-piece, I fitted a **90-degree ball valve**. This allows me to **shut off** the new pipework if needed.
3. **Connect the Pump** – The **pump intake pipe** is then connected to this outlet **after the isolation valve**.

Connecting the Return Line to the Top of the Tank

The second connection point is **at the top of the tank**, where hot water returns from the **solar collector**.

- My tank **did not have a spare top entry**, so I had to **modify the connection where the safety pressure relief valve was installed**.
- I installed a **T-piece** at this entry, allowing me to:
  - Screw the **safety valve** back into the system.
  - Plumb the **hot water return line** from the solar collector.

These two pipes—**cold water out and hot water return**—are typically **12.5mm copper pipes** running to the **rooftop collector**.

Safety Considerations: The Pressure Relief Valve

I was initially concerned that introducing very hot water near the safety valve might trigger it to release water, but in practice, this does not happen. The valve only activates when the water approaches boiling point.

Never Remove the Safety Pressure Valve!

This valve is critical for safety. If your thermostat malfunctions and stays in the on position, the water inside the tank could continue heating until it boils. This would create massive steam pressure, which, if unchecked, could cause a catastrophic explosion, destroying your house.

For proof, look up "steam explosion" on YouTube to see how dangerous this can be.

Why the Valve Releases Water Normally

During normal operation, it is expected that the pressure relief valve will occasionally release small amounts of water. This happens because:

- Water expands when heated, increasing pressure.
- Plumbing regulations prevent pressure from escaping back into the mains supply, as the mains one-way inlet valve blocks reverse flow.

I've noticed that on hot, sunny days, when the pump is running, the valve releases small amounts of water, which is completely normal.

Alternative Tank Connection Options

Depending on your specific hot water system, the connection points may vary. Here are alternative ways to plumb the system:

If Your Tank Has Extra Ports

- Some tanks come with unused top and bottom ports, usually, 1-inch BSP threaded.
- If available, simply remove the bungs and install:
    - A male-to-female adapter
    - A pipe connector

If You Can't Access the Bottom of the Tank

- Instead of connecting directly to the bottom, you can tap into the cold water inlet pipe feeding the tank.
- **Important**: This must be done after any valves or pressure release devices, ensuring proper function.

If You Can't Access the Top of the Tank

- If there's no spare port at the top, you can do what I did—T-piece into the safety valve connection.

For further clarification, I've added diagrams and photos at the end of this document to better illustrate the collector piping setup.

Choosing the Right Solar Collector

The key to a successful solar hot water system is selecting a collector that is both effective and cost-efficient.

Finding the Right Size and Type

- I use a 30-tube evacuated glass tube collector, which I purchased second-hand for $500 from local classifieds.
- Brand-new units sell for twice that price, so it's worth shopping around for a great deal.

Installation and Placement

- The collector must be mounted on the roof in the sunniest possible location.
- Ensure the plumbing is properly secured and the unit is firmly fastened to withstand strong winds and storms.
- The best angle for mounting is 90 degrees to the middle of the sunniest part of the sky for maximum solar exposure.

Sizing for Your Household

- A 30-tube collector is sufficient for two people.
- For larger households (e.g., five people) who take frequent showers, you may need:
    - A larger collector OR
    - Accept that you may need to rely on electric boosting occasionally.

Temperature Monitoring and Pump Control

This part of the project is where things get interesting. You'll need an electronic controller that monitors temperatures at two key points in your system:

1. The rooftop collector
2. The top of the hot water tank

How the Controller Works

- When the collector temperature rises significantly above the tank temperature, the pump turns on to transfer hot water down to the tank.
- In doing so, cooler water from the bottom of the tank circulates up to the collector to absorb more heat.
- Once the temperatures equalize, the pump switches off.
- This cycle repeats every 30 seconds to 3 minutes, depending on:
    - Distance between the collector and tank
    - Pump size
    - Differential temperature set point
    - Pipe insulation quality

Setting the Differential Temperature

- The differential temperature is the minimum temperature difference required between the collector and the tank before the pump activates.
- My system is adjustable for testing, and I've found that 12°C works best.
    - **Example**: If the tank temperature is 50°C, the pump will turn on when the collector reaches 62°C.

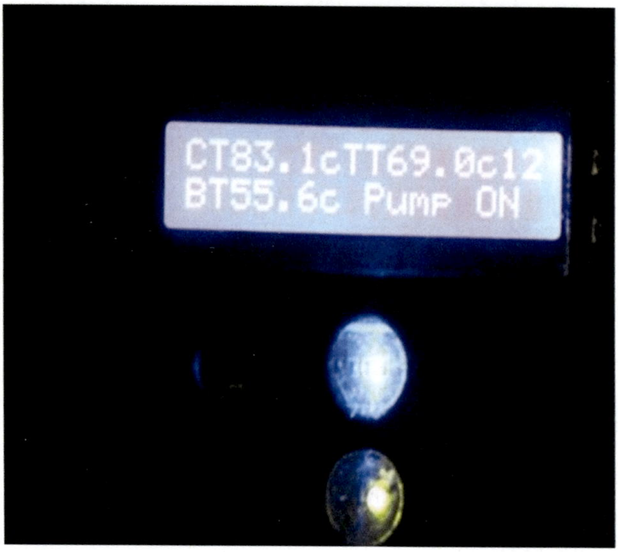

Insulating Your Pipes for Maximum Efficiency

Since you're moving very hot water, proper insulation is essential to prevent heat loss:

- **Hot water pipe (from collector to tank) → Must be well insulated**
    - I use double-layer insulation, which has noticeably improved efficiency.
- **Cold water pipe (from tank to collector) → Optional insulation**

- Keeping the cold pipe separate from the hot pipe prevents heat transfer.
- Although my tests show the cold pipe does warm up slightly, deciding whether to insulate it depends on cost-effectiveness for your system.
- I chose to insulate mine as part of my testing process.

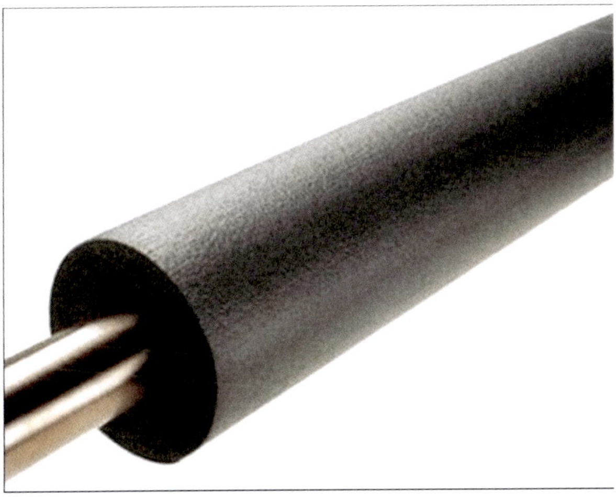

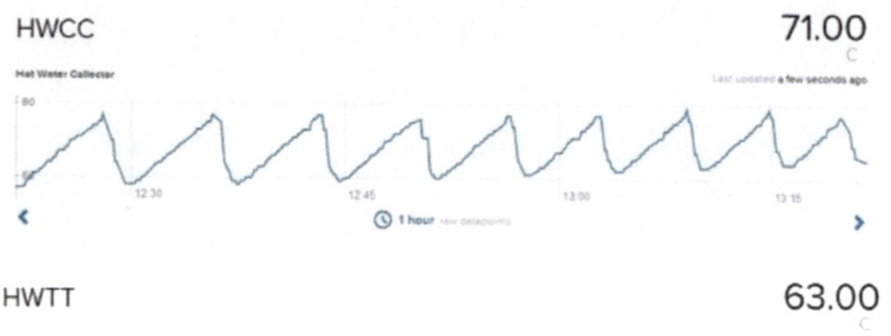

Collector Temperature Tracking Example

Here's a simple screen grab from my PC showing the temperatures of the Hot Water Collector (HWCC) and the Hot Water Top Tank (HWTT) during a typical October day.

- **HWCC (Hot Water Collector)**: You can see how the temperature rises and falls as the pump cycles on and off throughout the day, driven by sunlight exposure.
- **HWTT (Hot Water Top Tank)**: This shows how the water temperature in the tank adjusts as hot water is pumped in from the collector.

The pump activation occurs when the collector temperature exceeds the tank temperature by a set differential (12°C in my case), and deactivates once the temperatures equalize. This cycling helps to ensure the tank remains at an optimal temperature for use.

Here, you can see the tank has risen to 71Deg after 6 hours in the day sun.

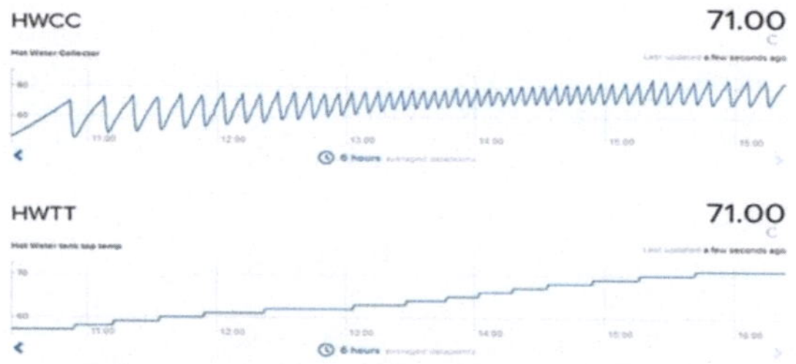

This data is provided by a micro-controller based system I have added to my installation which is not covered in this document.

Controlling the Pump

There are several ways to control the circulating pump in your solar hot water system. Below are a couple of options:

1. Simple Solar-Powered System (No Controller)

One of the simplest methods is to wire up a 12-volt photovoltaic solar panel directly to the pump. The size of the solar panel should match the power requirements of the pump (e.g., a 10-watt solar panel for a 10-watt pump).

How it works:

- When the sun is shining, the solar panel generates power, and the pump runs to circulate the hot water from the collector to the tank.
- When the sun stops shining, the pump stops, and the one-way valve ensures that the hot water stays in the tank.

While this system is simple, it has some drawbacks:

- **Inefficiency during cloudy weather**: The pump continues running during patchy cloud cover, leading to excess heat loss and unnecessary circulation.
- **Pump wear**: Since the pump runs constantly on sunny days, it can wear out faster.
- **Hot water left in the collector**: During periods of no sun, the water in the collector can stay too hot for extended periods, reducing efficiency.

Despite these issues, this method can work as a low-cost solution, but I found that using a properly programmed controller significantly improves system performance.

2. Using a Pre-existing or Second-hand Controller

A more efficient option is to use a controller designed for solar hot water systems. You can either purchase a new controller or find a second-hand unit from a decommissioned solar hot water system. This method allows for better control and optimization of the pump's operation, ensuring the system performs at its best.

The way I went, having a lot of electronics experience, is to build my own unit and have it programmable so I could try different ideas to make the system work well, and therefore save me money!

I chose an Arduino unit as these are very cheap and quick to attain via eBay (once again, eBay saves the day), and it was not a big jump for me to learn how to program it.

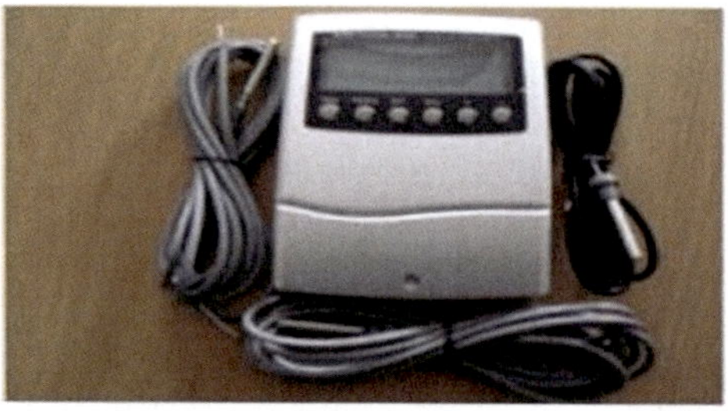

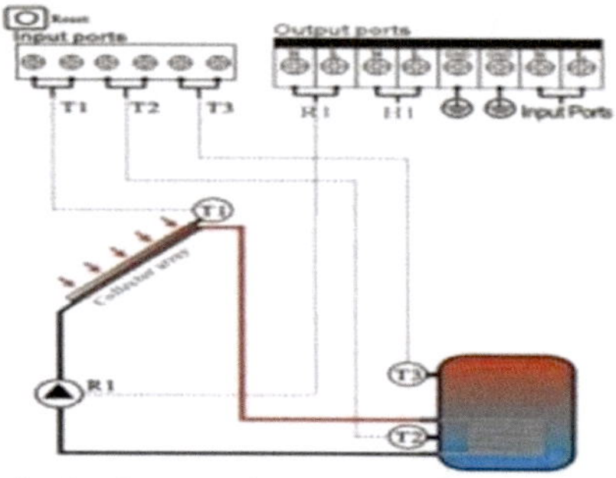

A typical aftermarket controller unit as seen on Ebay.

Controller Unit – The Arduino Solution

At the heart of the solar hot water system's control is the Arduino Uno, which provides the system with logic. An LCD screen is connected to this Arduino for easy monitoring of system temperatures and pump status.

Cost: When I built my system, the total cost for the Arduino and LCD screen was only $17, which is an affordable way to add smart control to your setup.

System Components:

- **Relay**: Controls the circulating pump.
- **Power Supply**: Provides the necessary power for the Arduino and the pump.

For the 12-volt power supply, you can use a simple wall pack or any power source that provides enough power to run both the Arduino unit and the circulating pump. However, I've taken it a step further by incorporating a 15-watt solar panel to power the pump. This solar panel is connected to a regulator and a battery, allowing the system to run off solar power when available. The Arduino unit is powered by a 9-volt plug pack (also called a "wall wart" in the USA).

By using this setup, I've made the system more self-sufficient, reducing reliance on grid power and enhancing the efficiency of the solar hot water system.

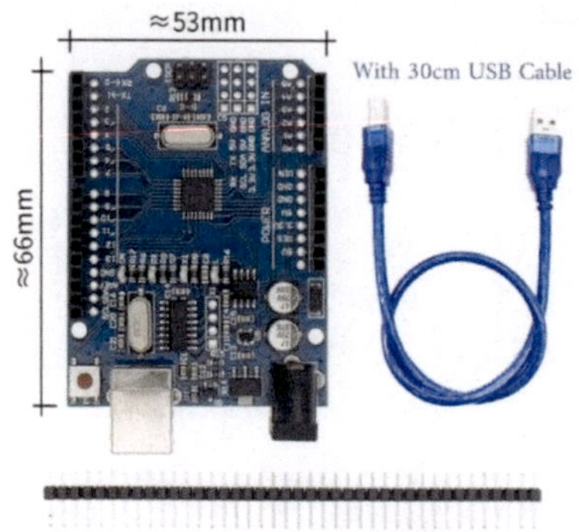
≈53mm
≈66mm
With 30cm USB Cable

UNO

LCD Keypad Shield
SELECT
LEFT
UP
DOWN
RIGHT
RST

Pump Speed Control and Noise Reduction

After running the system for a while, I found that the circulating pump was a bit noisy, so I added an electronic speed control to adjust the pump speed. I set it to about half speed, which reduced the noise without affecting the flow rate enough to compromise the system's performance.

How to Implement Pump Speed Control:

You can use a 1 Amp PWM 12-volt motor speed control circuit or kit, which is available from suppliers like Dick Smith Electronics, Jaycar, or eBay. These kits are affordable and easy to use. The setup is simple:

- Connect the 12V output from the control relay to the speed control unit.
- Use the variable resistor (pot) on the unit to adjust the pump speed to your liking, ensuring the noise level is manageable.
- Then, connect the output of the speed control unit to the pump input leads, ensuring all wiring is protected from weather and moisture.

I housed the speed control unit in the same box as the Arduino and LCD screen, which keeps the setup compact and organized.

Optional Redundancy for Pump Failure:

For added reliability, you could install two pumps side by side (in parallel), each with its own ball valve for isolation. This way, if one pump fails, you can quickly switch to the other pump and continue operation without disruption. You'd need a quick connector on the wiring to facilitate this.

While my pump, a TOPSFLO model from eBay ($80), has been running for almost 2 years without any issues, I believe that running the pump at a lower speed helps extend its lifespan.

However, if your pump is quiet enough, you might leave it running at full speed without issue.

Pump Failure and Overheating:

Be aware that if a pump fails, the solar collector on the roof can reach dangerously high temperatures, potentially damaging the glass tubes or causing the copper tubes to burst. Steam generation is also a risk. While I haven't had this issue myself, I would advise covering the collector with a tarp if the pump fails to prevent overheating until you can repair the system.

Safety Tip: Always prioritize your safety if working on the roof. Use a fall prevention harness or set up a barricade to avoid accidents.

System Programming:

I've included the programming code for the system, which you can use in your own project.

However, every installation is different, so you'll likely need to fine-tune it for your specific setup. You might have a larger pump, a different power supply, or other variations that will require adjustments.

Using DS18B20 Temperature Sensors with Arduino

The system utilizes three DS18B20 temperature sensors, each connected to the Arduino Uno, to monitor temperatures at key points in the system:

- **Collector Temperature**: To monitor the temperature of the solar collector.
- **Top Tank Temperature**: To monitor the temperature at the top of the hot water tank.
- **Bottom Tank Temperature**: To display the temperature at the bottom of the tank on the LCD screen.

The bottom tank sensor is mainly used for display purposes on the LCD, and the other two sensors are used to control the system's operation.

RTC (Real-Time Clock) Removal:

Initially, I used an RTC to lock out the pump from operating at night and to display the time. However, I found these features unnecessary, so I removed the RTC from the setup. The RTC code is still included in the program for reference, and you can choose to use it if desired. If you're not using the RTC, simply delete any lines that refer to it, such as:

- #include "RTClib.h"
- //RTC_DS1307 RTC;
- //RTC.adjust(DateTime(__DATE__, __TIME__));

If you do remove the RTC, it will not impact the performance or resources of the system.

Setting Up the Arduino and LCD:

To set up the unit, you will need:

1. Arduino Uno and 16x2 LCD screen shield.

2. Arduino IDE: Download the free program from the Arduino website to write and upload code to the Arduino.
3. Libraries: Make sure you have the correct libraries installed:
    - Wire
    - LCD4Bit_mod
    - stdio
    - stdlib

Some libraries are pre-installed with the Arduino IDE, but if your LCD screen is different, you may need to adjust the wiring and install a different library.

Button Controls:

My LCD unit has five buttons, with two programmed to adjust the differential setting:

- **Up Button**: Increases the differential by 1 degree.
- **Down Button**: Decreases the differential by 1 degree.

This allows for easy adjustment without needing to reprogram the unit. However, keep in mind that resetting the unit will revert to the default setting in the code (e.g., int diff = 10; for the differential setting and int pumpoffdelay = 40; for pump-off delay).

Wiring and DS18B20 Setup:

The DS18B20 temperature sensors are connected using Cat 5 twisted-pair cable. Each sensor requires three wires:

- +5V (Vdd)
- Ground (GND)
- Data Line (DQ)

You'll also need a 4.7kΩ resistor between the data line (DQ) and the 5V line at the sensor end to pull the data line high for digital communication. The digital signal ensures accurate, interference-free temperature readings.

The DS18B20 is highly accurate, with a resolution of 0.06°C, and can provide temperature readings with 1 decimal place of accuracy, which is sufficient for our Hot Water System (HWS) control needs.

Maxim 18B20 temperature sensor diagram.

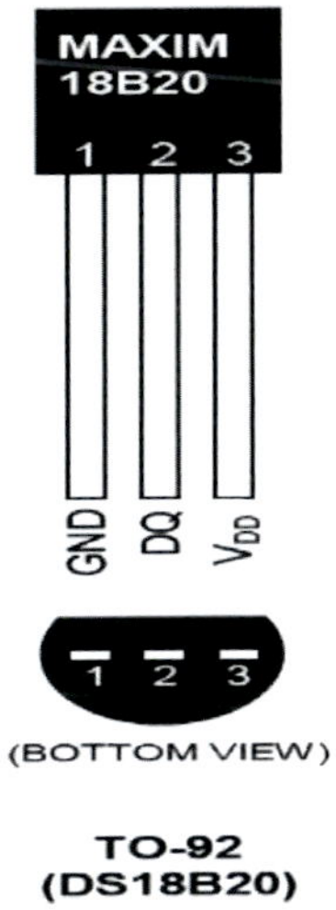

A pre built temperature sensor with water proof casing and cable

Wiring Up and Installing Temperature Sensors

In my project, I built custom waterproof temperature sensor units using DS18B20 sensors. However, you can also purchase pre-made waterproof DS18B20 sensors for ease.

Important: Do not attempt to install temperature sensors directly into the hot water of the tank. Doing so would require breaking the tank's "seal," which could be hazardous and may compromise the system's integrity.

Sensor Installation Methods:

1. Collector Tank:

My collector tank had a recessed hole designed specifically for a temperature probe. Fortunately, my homemade sensor unit fit into this hole perfectly. If your collector does not have a designated hole, you can simply bond the temperature sensor with glue to the outside of the tank. This will provide an accurate temperature reading.

2. Hot Water Tank:

For the water tank, you have several installation options:

- **Under the Insulation**: You can place the sensor under the tank's insulation, ensuring the probe is pressing against the side of the steel tank. This is a safe and easy method.
- **Accessing the Insulation**: To position the sensor correctly, you may need to:
  - Open the top cover of the tank.
  - Remove some screws or cut a small hole through the outside of the tank to access the insulation.
  - Do NOT cut into the tank itself! Cutting into the water tank could compromise the system's integrity and safety.

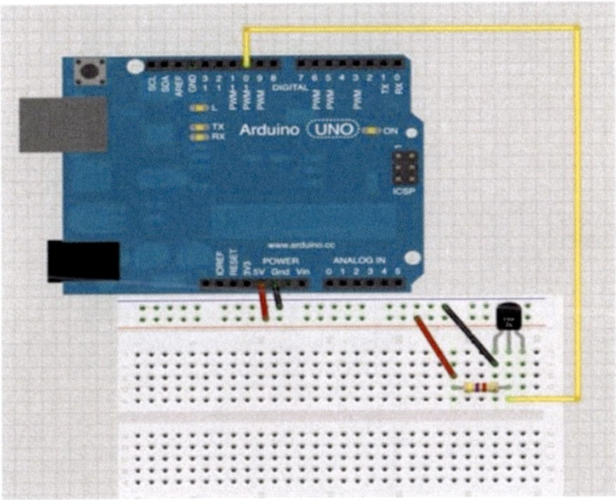

Example test diagram

Useful Resources:

For detailed wiring instructions and reference, visit the Arduino website for DS18B20 sensor information.

http://playground.arduino.cc/Learning/OneWire

If the link is no longer available, simply search for Arduino DS1820.

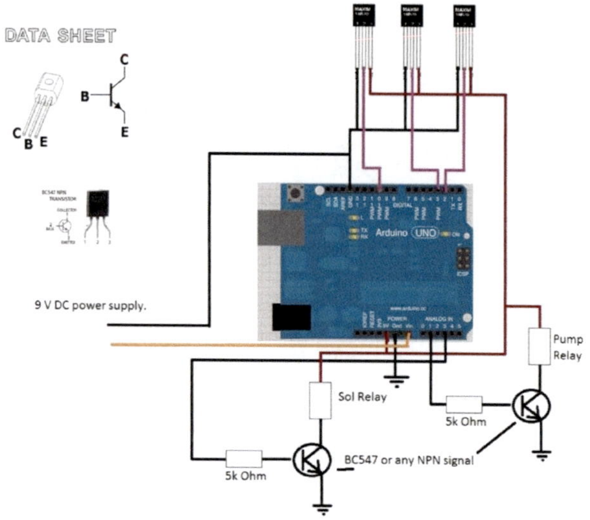

Wiring and Configuration

In this setup, I'll describe how to wire your system for temperature sensing, LCD display, and pump control. If you're using a different LCD screen or components, keep in mind that pin assignments might differ. You may need to change the pin numbers in the code accordingly.

LCD Wiring:

If you're using an LCD screen like mine, refer to the wiring diagram provided. However, your LCD may use different pins. Ensure that the LCD doesn't overlap with the pins used by the temperature sensors or A1/A3 for the relays. For example, if your system uses A1 for the pump control, you could modify it to A2 or A5, ensuring no conflicts in the wiring.

Dump Solenoid (Overheat Protection):

To enhance safety, I have added a dump solenoid to my system. This is an additional feature that releases excess hot water from the top of the tank if the temperature reaches a critical level, which is safer than relying solely on the safety pressure valve.

Why a Dump Solenoid? On occasion, my system has reached temperatures as high as 85°C, which is too hot for comfort and could cause burns. Therefore, I've added the solenoid as an extra layer of protection. The dump solenoid is programmed to release water from the top of the tank for 60 seconds if the water temperature exceeds the set point.

Code Snippet:

In the code, you'll find the following line:

```
const int tmax = 75;  // Max top tank temperature set point for water solenoid to dump excess hot water
```

This value, 75°C, is the threshold at which the dump solenoid is triggered. You can adjust this value based on your preferences and system's characteristics.

Safety Considerations:

It's crucial to set up your overheat protection correctly. A dump solenoid can prevent dangerous overheating, especially in systems that may occasionally overheat beyond safe limits.

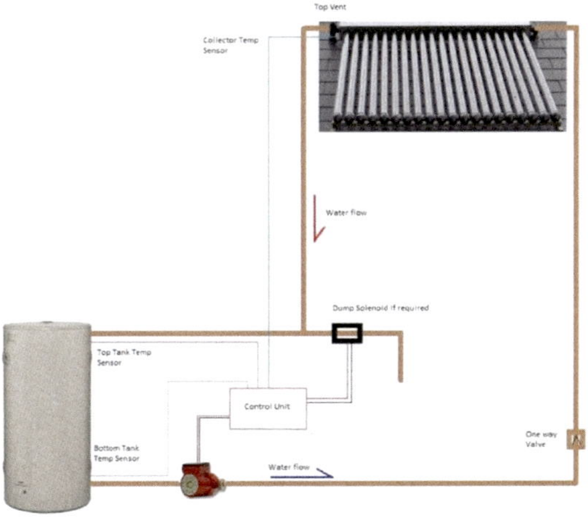

I've recently updated my unit's programming to include a "mixing" function to prevent dumping valuable heat on hot days. The idea is

simple: if the top tank temperature reaches something like 80°C, while the water at the bottom might still be only 45-55°C, the system will run the pump for an extra 3 minutes during each cycle to mix the water. This helps to distribute the heat evenly throughout the tank.

If, after this extra mixing time, the temperatures still get too high (e.g., late in the day), the dump solenoid kicks in to release the excess heat. However, after 12 hours, the water in the tank begins to stratify. This means the top stays hot, but the bottom cools significantly. You might see 60°C at the top and only 25°C at the bottom in the mornings.

In addition to this, I've added frost protection. If the collector temperature drops to a dangerously low level (around 2°C), the system will automatically turn on the pump for 30 seconds, circulating water from the bottom of the tank to prevent the collector and pipes from freezing. This prevents ice from forming inside the system, which could otherwise lead to damage due to expanding ice.

And here is the CODE! Good luck!

This is tested and working code, copied from this document back into the microcontroller to prove it has no mistakes.

```
/*ReadDS18B20three

Reading three DS18B20s seperately, keeping track of real time,

displaying to ekitszone 16x2 LCD shield,

and controlling a pump if collector temperature is more than 7 deg C

above bottom tank temperature and if tank is not above 95deg c.

RTC connected to A4 and A5

also transmitting two temperatures int values via 433 mhz radio to outside reciever
using virtualwire

 */

#include <Wire.h>
```

```
#include "RTClib.h"

#include <LCD4Bit_mod.h>

#include <stdio.h>

#include <stdlib.h>

#include <VirtualWire.h>

LCD4Bit_mod lcd = LCD4Bit_mod(2);

#define tture1 2  // data line of DS18B20 sensor connected to D2

#define tture2 3

#define tture3 10

//RTC_DS1307 RTC;

char Sensor1CharMsg[6];

long int Sensor1Data;

int pumpon, solon, ct, tt, bt, x, h, m, w, d, adc_key_in; //initiate intergers,

//for pump state, ct temp and bt temp, and data from sensors, pushbuttons, and
timing constants.

//int tankmax = 95;  // set maximin temp of top tank sensor

int diff = 12;  // initial differential, temp above bt temp that pump turns on

int pumpoffdelay = 25;

int frost = 2;  // frost set point for frost protection

int HighByte, LowByte, TReading, SignBit, Tc_100, Whole, Ttemp, Fract, at;

const int pump = A1;  //define pin for pump relay

const int pumpLED = A2;     // define pin for LED output

const int sol = A3;  // define pin for overtemp water release solenoid

const int tmax = 80;  //max top tank temp set point for water solenoid to dump
excess hot water void setup() {

  Wire.begin();

  //RTC.begin();

  // following line sets the RTC to the date & time this sketch was compiled

  //RTC.adjust(DateTime(__DATE__, __TIME__));
```

```
  pinMode (pump, OUTPUT);  //set pins to output pump relay

  pinMode (pumpLED, OUTPUT); // sets LED pin to show pump is on

  pinMode (sol, OUTPUT);  // sets pin to output solenoid

  lcd.init();     //initiate display and setup standing display elements

  setupscreen();

  //For each tture sensor: Do a pinMode and a digitalWrite

  pinMode(tture1, INPUT);

  pinMode(tture2, INPUT);

  pinMode(tture3, INPUT);

  digitalWrite(tture1, LOW);//Disable internal pull-up.

  digitalWrite(tture2, LOW);

  digitalWrite(tture3, LOW);

  Serial.begin(9600);

  delay(100);//Wait for newly restarted system to stabilize

  // VirtualWire setup

  vw_setup(500); // Bits per sec

}

void loop(){

  adc_key_in = analogRead(0);

  Sensor1Data = (tt*100)+ct;//round off top tank temperature into a whole int and
send it to first two digits of the four digit string

  itoa(Sensor1Data,Sensor1CharMsg,10); // send four digit string containing both
temps

  w = w +1;

  //Serial.println(adc_key_in);    //was used for setting up buttons

  if (adc_key_in < 150) {   //if "UP" button is pressed, increment diff value up one.

    diff = diff + 1;

    delay(50);
```

```
}

if ((adc_key_in > 300) && (adc_key_in < 350)) {    //if "DOWN" button is
pressed, increment diff value down one.

  if (diff > 3) {         // diff value can not go below 3 (coz that'd be daft).

    diff = diff - 1;

    delay(50);

  }

}

if (diff < 10)  // this is just to keep the display neat if the diff is reduced back
below 10

{

  lcd.cursorTo(1, 15);

  lcd.println(" ");

}

 delay(200);

readTture(tture1);//N.B.: Values passed back in globals

Serial.print("ct");

ct = Whole;  // ct is collector temperature, used for pump operation calculations


lcd.cursorTo(1, 2);  //top line cursor position 3

printTture();//N.B.: Takes values from globals. Also...

Serial.print("   ");


delay(200);// Delay... must not be too short.

lcd.cursorTo(1, 9);

Serial.print("tt");

readTture(tture2);//Now read and report 2nd tture.

tt = Whole;  // store this temperature value as current top tank temp   printTture();
```

```
delay(200);//

Serial.print("  bt");//Start new line

lcd.cursorTo(2, 2);  //line=2, x=0

readTture(tture3);//N.B.: Values passed back in globals

bt = Whole;  //bt is bottom tank temperature used for pump operation
calculations

printTture();//N.B.: Takes values from globals. Also...

Serial.print("  ");


 if (pumpon == true) {

   Serial.print("PUMP ON");

 }

else {

   Serial.print("PUMP OFF");

 }

 if (solon == true) {

   Serial.print(" SOL ON");

}

else {

   Serial.print(" SOL OFF");

}

Serial.print("  pump off delay is ");

Serial.print(pumpoffdelay);

Serial.print(" diff ");

Serial.print(diff);

//Serial.print("  "Ttemp);

Serial.print("\n");
```

```
// DEBUG radio
//Serial.print("Sensor1 Integer: ");
//Serial.print(Sensor1Data);
//Serial.print(" Sensor1 CharMsg: ");
//Serial.print(Sensor1CharMsg);
//Serial.println(" ");
//Serial.print(hundreds);
//Serial.print("  ");


delay(100);// Delay... must not be too short.
x = x +1;
if (x == 10) {
  vw_send((uint8_t *)Sensor1CharMsg, strlen(Sensor1CharMsg)); //send string to
radio
  vw_wait_tx(); // Wait until the whole message is gone
  //Serial.println(Sensor1Data);
  Serial.print("sent ");
  Serial.println(Sensor1CharMsg);
  if (tt >= tmax) {  //turns on heat dump solenoid if temp of tank is above tmax
deg c       solon = true;
  }
  else {
    solon = false;
  }
  delay(1000);
  setupscreen();
  x = 0;
}
if ((ct > (tt + diff)) || (ct < frost))  { // if its hot enough to run, or its frosty
```

```
  pumpon = true;
  w = 0;
  }
  else {
  if (w == pumpoffdelay) {   // this ensures that if the pump turns on, it STAYS on for 20 cycles (seconds)
    pumpon = false;
    w = 0;
  }
}
if (tt == (tmax - 2)) {   // tank "mixing" feature if it gets very warm (close to tank max temp cut off point, where it will dump hot water
 pumpoffdelay = 200;
}
else {
 pumpoffdelay = 25;
}
 if (w == 101) {
  w = 0;
 }
 if (pumpon == false) {
   lcd.cursorTo(2, 13);
   lcd.println("OFF");
   digitalWrite (pump, LOW);
  digitalWrite (pumpLED, LOW);
 }
 if (pumpon == true) {
   lcd.cursorTo(2, 13);
   lcd.println("ON ");
```

```
    digitalWrite (pump, HIGH);
   digitalWrite (pumpLED, HIGH);
  }
  if (solon == true) {
    digitalWrite (sol, HIGH);
  }
  if (solon == false) {
    digitalWrite (sol, LOW);
  }
}
void OneWireReset(int Pin) // reset.  Should improve to act as a presence pulse
{
  digitalWrite(Pin, LOW);
  pinMode(Pin, OUTPUT); // bring low for 500 us
  delayMicroseconds(400);
  pinMode(Pin, INPUT);
  delayMicroseconds(400);
 }//end OneWireReset
void OneWireOutByte(int Pin, byte d) // output byte d (least sig bit first).
{
  byte n;
  for(n=8; n!=0; n--)
   {
    if ((d & 0x01) == 1)  // test least sig bit
    {
      digitalWrite(Pin, LOW);
      pinMode(Pin, OUTPUT);
      delayMicroseconds(5);
```

```
    pinMode(Pin, INPUT);
    delayMicroseconds(60);
   }
   else
   {
    digitalWrite(Pin, LOW);
    pinMode(Pin, OUTPUT);
    delayMicroseconds(60);
    pinMode(Pin, INPUT);
   }
   d=d>>1; // now the next bit is in the least sig bit position.
  }
}//end OneWireOutByte
byte OneWireInByte(int Pin) // read byte, least sig byte first
{
  byte d, n, b;
  for (n=0; n<8; n++)
  {
   digitalWrite(Pin, LOW);
   pinMode(Pin, OUTPUT);
   delayMicroseconds(5);
   pinMode(Pin, INPUT);
   delayMicroseconds(5);
   b = digitalRead(Pin);
   delayMicroseconds(50);
   d = (d >> 1) | (b<<7); // shift d to right and insert b in most sig bit position
  }
  return(d);
```

```
}//end OneWireInByte

void readTture(byte Pin){

 //Pass WHICH pin you want to read in "Pin"

 //Returns values in... (See global declarations)

 OneWireReset(Pin);

 OneWireOutByte(Pin, 0xcc);

 OneWireOutByte(Pin, 0x44); // perform temperature conversion, strong pull up
for one sec

 OneWireReset(Pin);

 OneWireOutByte(Pin, 0xcc);

 OneWireOutByte(Pin, 0xbe);

 LowByte = OneWireInByte(Pin);

 HighByte = OneWireInByte(Pin);

 TReading = (HighByte << 8) + LowByte;

 SignBit = TReading & 0x8000;  // test most sig bit

 if (SignBit) // negative

 {

  TReading = (TReading ^ 0xffff) + 1; // 2's comp

 }

 Tc_100 = (6 * TReading) + TReading / 4;    // multiply by (100 * 0.0625) or 6.25

 Whole = Tc_100 / 100;  // separate off the whole and fractional portions

 Fract = Tc_100 % 100;

};//end readTture

void printTture(){//Uses values from global variables.

 //See global declarations.

 //N.B.: No new line inside printTture

 if (SignBit) // If it's negative

 {
```

```
  Serial.print("-");
 };
 Serial.print(Whole);
 Serial.print(".");
 if (Fract < 10)
 {
  Serial.print("0");
 };
 Serial.print(Fract);
 char temp [33];
 char D [33];
 char Dec [10];
 itoa (Whole,temp,10);  //convert the interger value Whole to a char value
 // and call it temp to display to the lcd as it cannot display intergers
 itoa (diff,D,10);
 d = floor (Fract/10);
 itoa (d,Dec,10);
 lcd.printIn(temp);  //Note, wont print "-" negative to lcd screen
 lcd.printIn(".");
 lcd.printIn(Dec);
 lcd.cursorTo(1, 14);
 if (diff<10) {
   lcd.printIn("0");  //created a strobing "diff" value. Need to work out why.
 }
 lcd.printIn(D);
}
void setupscreen(){
 lcd.clear();    // clears display
```

```
  lcd.println("CT    c");
  lcd.cursorTo(2, 0);
  lcd.println("BT    c");
  lcd.cursorTo(1, 7);
  lcd.println("TT    c");
  lcd.cursorTo(2, 8);
  lcd.println("Pump");
};
```

This is a typical ball valve (or isolating valve) one could use to close off the system top stop water flowing up into the collector

This photo is to show how to connect in the bottom collector pipe (cold pipe) if you have no other access points on your hot water tank.

Originally there was only the vertical pipe, going into the tank via a 90 deg elbow. The elbow was removed and the T piece put in its place. This extra horizontal connection is where the isolation valve (blue handle) was put in and this pipe then leads away to the circulating pump and then on up to the collector. This way when the pump starts, it draws in cool water from the bottom of the tank through the pump and up to the collector to collect heat from the sun.

This photo shows the top of the tank showing the plumbing and the modification. The black insulated pipe on the left is the HWS outlet pipe. This is where hot water is fed to the hot water taps in the house. This is the connection you can NOT use to introduce hot water from the collector into the system. The connection on the right is where the safety valve was originally plumbed in. I have removed insulation on this for the photo. The bare copper pipe is the drain for the valve, which is open to the atmosphere at the drain below the tank. The safety valve was removed and a nipple and a T piece was plumbed in as shown so that an extra pipe could be connected in. Hot water from the collector on the roof, then flows into the tank (past the safety valve sensor) into the tank. This should not affect the operation of the safety valve. If it does then either the differential value in the settings in the controller module is set

too high, or the temperature sensor is reading inaccurately, or the valve is faulty (passing as lower temperatures).

*** Important update regarding Speed controllers on pumps. If you are using a speed controller to slow down the speed of the pump to control pump noise, ensure you have connected across the output of the controller a smoothing capacitor (such as a 2200uF 25v) and ensure it is connected the correct way around. If you fail to do this you may damage the electronics in the pump itself particularly if it is one of the Topsflo type of pump which has its own brushless motor control electronics inside the housing. Your pump should run smooth and continuously. If it runs erratically or stops occasionally you may have a problem in this area.

Disclaimer:

I accept no responsibility for any damage or injury to persons or property resulting from the use of this guide. This guide is intended for individuals with some DIY skills and a basic understanding of plumbing and electrical systems.

Important:

If you are working with mains power (240V), you must be a licensed electrician. If you do not have experience with plumbing, please consult a certified plumber before proceeding, as it may be illegal in your area to perform plumbing work without certification. In some regions, “repairs” may be allowed, but “new installations” may require professional certification.

Please ensure you comply with your local regulations regarding one-way valves, tempering valves, and other safety measures. Do not remove your safety valve, nor rely on it to handle excessive heat in the solar hot water system. Always follow safety protocols to prevent overheating issues.

Made in United States
Orlando, FL
05 November 2026

90230151R00033